Travel Adventures

The Great Barrier Reef

Place Value

Dona Herweck Rice

Consultants

Colene Van Brunt
Math Coach
Hillsborough County Public Schools

Publishing Credits

Rachelle Cracchiolo, M.S.Ed., *Publisher*
Conni Medina, M.A.Ed., *Managing Editor*
Dona Herweck Rice, *Series Developer*
Emily R. Smith, M.A.Ed., *Series Developer*
Diana Kenney, M.A.Ed., NBCT, *Content Director*
June Kikuchi, *Content Director*
Susan Daddis, M.A.Ed., *Editor*
Stacy Monsman, M.A., *Editor*
Kevin Panter, *Senior Graphic Designer*

Image Credits: pp.10–11 Doug Perrine/Getty Images; pp.12–13 Jeff Hunter/Getty Images; all other images from iStock and/or Shutterstock.

Library of Congress Cataloging-in-Publication Data

Names: Rice, Dona, author.
Title: Travel adventures : the Great Barrier Reef / Dona Herweck Rice.
Other titles: Great Barrier Reef
Description: Huntington Beach, CA : Teacher Created Materials, 2019. | Includes index. | Audience: K to Grade 3. |
Identifiers: LCCN 2017054931 (print) | LCCN 2018009298 (ebook) | ISBN 9781480759718 (eBook) | ISBN 9781425856779 (pbk.)
Subjects: LCSH: Great Barrier Reef (Qld.)--Juvenile literature. | Coral reef ecology--Juvenile literature.
Classification: LCC QE566.G7 (ebook) | LCC QE566.G7 R53 2019 (print) | DDC 577.7/89--dc23
LC record available at https://lccn.loc.gov/2017054931

Teacher Created Materials
5301 Oceanus Drive
Huntington Beach, CA 92649-1030
www.tcmpub.com

ISBN 978-1-4258-5677-9

Printed in China
51497

Table of Contents

Nature's Playground

The Great Barrier Reef is big. It is beautiful. It is very old. And it is alive!

Great Barrier Reef

It is made of smaller **coral reefs**. Many sea **creatures** live there. Some are part of the reef.

Let's Do Math!

Imagine you see 16 clown fish at the Great Barrier Reef. Draw or place objects on ten frames to answer the questions.

1. Are there enough clown fish to make a ten?

2. Are there any leftover clown fish after you make a ten? If so, how many?

What Is a Reef?

A reef is a bar of rock and coral. A reef lies just under the water in the ocean.

Living coral grows on the rock. The rock is made from coral that has died.

Live coral is colorful.

Sea Creatures

There is a lot of life in the Great Barrier Reef! Sea creatures of all kinds call it home.

A green sea turtle and some fish swim near the reef.

Hundreds of types of coral live there. Fish of all colors live there. Dolphins and rays live there, too.

coral trout

Let's Do Math!

You are swimming with your brother at the reef. You see 22 blue tangs. Your brother sees 24 blue tangs. Use the phrases "**greater** than" and "**less** than" to compare the number of fish.

1. 22 is ______ 24
2. 24 is ______ 22

spotted eagle ray

The reef is also home to many types of birds. They live on its **islands**.

red-tailed tropicbird

silver gull

Take Care!

Many people visit the Great Barrier Reef. They must help take care of it. We want the reef to last!

Let's Do Math!

You **snorkel** through the Great Barrier Reef. You see 12 blue triggerfish. You see 20 green parrotfish. You see 13 yellow cuttlefish. Use > or < to compare the number of fish.

1. 12 ______ 20

2. 20 ______ 13

3. 13 ______ 12

Problem Solving

You are ready to snorkel at the Great Barrier Reef. Down you go! You see...

- 12 rays
- 21 blue tangs

1. How many groups of ten rays are there? Are there any leftover rays? If so, how many?

2. How many groups of ten blue tangs are there? Are there any leftover blue tangs? If so, how many?

3. Use >, <, or = to compare the number of rays and blue tangs.

   12 ______ 21

Glossary

coral reefs—hard materials that are formed on the bottom of the sea

creatures—animals

greater—more

islands—areas of land surrounded by water

less—fewer

snorkel—to swim underwater with a breathing tube

Index

Answer Key

Let's Do Math!

page 7:

1. yes
2. yes; 6 leftover clown fish

page 15:

1. less than
2. greater than

page 19:

1. <
2. >
3. >

Problem Solving

1. 1 group of ten; yes; 2 leftover rays
2. 2 groups of tens; yes; 1 leftover ray
3. <